PERSISTENT POISONS

PERSISTENT POISONS

Chemical Pollutants in the Environment

Mary-Jane Schneider, Ph.D.

Ill fares the land,
to hast'ning ills a prey
—Oliver Goldsmith
The Deserted Village (1770)

The New York Academy of Sciences
New York, New York
1979

Copyright © 1979 by The New York Academy of Sciences. All rights reserved.
Except for brief quotations by reviewers, reproduction of this publication in
whole or in part by any means whatever is strictly prohibited without written
permission from the publisher.

Second printing, September, 1980.

Library of Congress Cataloging in Publication Data

Schneider, Mary-Jane, 1939-
 Persistent poisons.

 Based on Health effects of halogenated aromatic
hydrocarbons, the proceedings of a conference held in
New York in June, 1978, by the New York Academy of
Sciences, which was published as vol. 320 of the Annals
of the New York Academy of Sciences.
  1. Halocarbons—Toxicology. 2. Aromatic compounds—
Toxicology. 3. Halocarbons—Environmental aspects.
4. Aromatic compounds—Environmental aspects.
I. New York Academy of Sciences. II. Title.
RA1242.H35S36   615.9'51'1   79-13403

The New York Academy of Sciences believes that it has a responsibility to
encourage the dissemination of scientific information. For this reason, *Persistent
Poisons: Chemical Pollutants in the Environment*, was written for the general
public on the basis of the proceedings of an Academy conference entitled Health
Effects of Halogenated Aromatic Hydrocarbons. The full proceedings of the
conference have been published as Volume 320 of the *Annals of The New York
Academy of Sciences*. Positions taken by the conference participants whose
work forms the basis for this volume are their own and not those of The
Academy. The Academy has no intent to influence legislation by the publishing
of such opinions.

PCP
Printed in the United States of America
ISBN 0-89766-023-4

Contents

	Page
Acknowledgments	vi
PART I—INTRODUCTION	1
PART II—THE HALOGENATED AROMATIC HYDROCARBONS	9
Chapter 1—DDT and Other Organochlorine Pesticides	11
Chapter 2—PCBs	13
Chapter 3—Dioxins and Dibenzofurans; Chlorophenols	18
Chapter 4—PBBs	23
Chapter 5—Other Pollutants	25
PART III—HEALTH EFFECTS	27
Chapter 6—General Symptoms	29
Chapter 7—The Search for Underlying Metabolic Disturbances	35
Chapter 8—Neurologic Effects	39
Chapter 9—Immunologic Effects	44
Chapter 10—Carcinogenicity	47
Chapter 11—Reproductive Effects	50
PART IV—TAKING ACTION	57
Chapter 12—Decontamination	58
Chapter 13—Identifying the Problems	62
Chapter 14—Avoiding Future Contamination	65

Acknowledgments

This book is based primarily on a conference held by The New York Academy of Sciences in June of 1978 entitled International Conference on Health Effects of Halogenated Aromatic Hydrocarbons, part of a Science Week, the general title of which was The Scientific Basis for the Public Control of Environmental Health Hazards. The conference was chaired by Dr. William J. Nicholson of the Mount Sinai School of Medicine of the City University of New York and Dr. John A. Moore of the National Institute of Environmental Health Sciences. Various persons reviewed the original manuscript and made valuable suggestions. Chief among them were Dr. Alastair Hay, Ms. Judith Leire, Dr. David P. Rall, Dr. Dora Wassermann, and Dr. Marcus Wassermann.

Financial support for the writing of this book was provided in part by the National Institute of Environmental Health Sciences.

Bill M. Boland
Executive Editor

PERSISTENT POISONS

Part I
INTRODUCTION

Séveso

On July 10, 1976, an explosion in a chemical factory near Milan, Italy sent a cloud of poison gases over the suburban town of Séveso. Among the chemicals that settled on the houses and gardens of the region was dioxin, or TCDD, a highly toxic by-product in the manufacture of herbicides and disinfectant soaps. Several thousand persons lived in the contaminated region.

Authorities at first minimized the importance of the accident. Only after domestic and wild animals began to die, and children were brought to the hospital with chemical burns, were plans made to move people from the contaminated area. Evacuation was not completed until over three weeks after the accident.

Within the next few months, some of the children began to develop a distinctive skin rash. Medical surveys began to detect signs of damage in the blood, liver, nerves, and vision of some of the exposed people. Because of concern over the possibility of birth defects, about 100 women underwent legal or illegal abortions.

Three years after the accident, a 215-acre area remains sealed behind yellow plastic fences. Carcasses of animals, plastic bags of vegetation, and piles of topsoil removed from the less highly contaminated areas await decision on how to dispose of them. Over 200 persons still live in temporary quarters, uncertain whether they will ever be able to return to their homes. The Italian regional government has filed a 113 billion lire * claim against Givaudan, a subsidiary of Hoffmann-La Roche, to pay for the clean-up and health surveillance program.

Fortunately, no serious illnesses or deaths have been traced to the accident, but the most serious consequences of the "Italian Hiroshima" may be yet to come. Dioxin lingers indefinitely in the body and continues to cause damage. Long-term effects, including cancer, may not show up for many years.

---

* Approximately 135 million dollars.

Michigan

At the time of the Séveso explosion, an American medical team was preparing to launch an investigation into another accident that had occurred three years earlier in the state of Michigan. There had been no explosion to announce the advent of this disaster. It began as a mysterious ailment that struck whole herds of dairy cattle in the fall of 1973. The animals lost their appetites, grew thin and weak, developed open sores that would not heal, produced dwindling quantities of milk, and became sterile or gave birth to dead calves. Months of detective work by a few tenacious investigators finally produced the explanation: toxic chemicals had found their way into the cattle feed. Not only were the chemicals making the cows sick, they were being excreted in the milk and contaminating Michigan's entire food supply.

The chemicals were polybrominated biphenyls, or PBBs, ingredients of a fire retardant manufactured by Michigan Chemical Corporation and trade-named Firemaster®. Michigan Chemical also produced magnesium oxide, called Nutrimaster®, used as a feed supplement for dairy cattle. Through a mix-up at the factory, Firemaster rather than Nutrimaster had been shipped to the farm cooperative that mixed dairy feed and distributed it to farmers around the state.

By May 1974, ten or eleven months after the original mix-up had probably occurred, the Michigan Department of Agriculture quarantined dairy farms where sick cows were producing PBB-contaminated milk. Field tests during the next few months found PBBs in the milk of healthy cows and in butter, cheese, and beef. By 1976, 29,800 of Michigan's finest dairy cattle had died or been destroyed. Because feed-mixing procedures at the farm cooperative caused contamination of other animal feeds, an additional 1,470 sheep, 5,920 pigs, and about 1.5 million chickens had to be slaughtered.

For over a year before the accident was discovered and subsequently while authorities debated what action to take, thousands of Michigan residents had been consuming milk, meat, and eggs that contained significant amounts of PBBs. Farm families, who had consumed the largest quantities of the chemicals, began to complain of headaches, fatigue, joint pain, and numbness in fingers and toes.

An early survey by the Michigan Department of Public Health found these symptoms to be equally common among a control group supposedly unexposed to the chemicals, and concluded that the PBBs were not causing significant health problems. But these conclusions were questioned when a later investigation, by a team from the Mt. Sinai School of Medicine in New York City, discovered that virtually everyone in Michigan had significant amounts of PBBs in their blood and body tissues and that, in fact, Michigan residents were suffering poorer health than a comparable but unexposed group of Wisconsin residents.

Little is known about the effects of exposure to PBBs, since few people have previously been in contact with the chemicals. Like the dioxin of Séveso, they seem to cause vague symptoms, none of them recognizably severe. But in Michigan as in Séveso, the long-term consequences may be yet to appear.

Yusho

A hint at what the future might bring to Séveso and Michigan comes from a third chemical disaster that occurred several years earlier in Japan. This one began insidiously, with a three-year-old girl, brought to a Kyushu University hospital in June 1968 with a skin rash. Soon others, including the child's parents and elder sister, appeared at the clinic with a similar rash, complaining also of swelling and discharge from the eyes, dark brown pigmentation of the skin and nails, headaches, and feelings of weakness. The disease spread through Western Japan and eventually struck some 1200 people.

Scientists painstakingly traced the problem to a specific batch of rice oil—manufactured in February 1968 at a specific factory—which was used for cooking by all the affected families. The oil was found to be heavily contaminated with heat-exchanger fluid that had leaked into the oil during processing.

PCBs, the major ingredients of the contaminating fluid, had long been known to produce rashes and other skin symptoms in workmen, and doctors at first concentrated on treating these problems. The more general complaints were not considered significant. With the passage of time, however, the obvious skin problems have tended to disappear, while the general symptoms persist and grow worse. Laboratory tests have shown disturbances in the liver, blood, nerves, immune response, and reproductive function of some patients. PCBs can still be measured in their blood and tissues. There is some indication that the cancer rate may be unusually high among these people, although it is still too soon after the accident to be certain.

At the time of the Yusho incident PCBs were relatively obscure industrial chemicals, only recently discovered to be common environmental pollutants. Since then we have learned that virtually all of us carry small amounts of the substances within our bodies. Furthermore, PCBs are close relatives of PBBs, dioxin, and other chemicals to which we are increasingly being exposed.

The "Yusho" patients—Yusho is Japanese for "oil disease" —constitute the largest group of people to have suffered from PCB poisoning. Their symptoms and bodily processes continue to be carefully monitored in search of clues as to the possible health effects the rest of us may expect from a lower but long-term exposure to such chemicals.

• • •

The dioxin of Séveso, the PBBs of Michigan, and the PCBs of Japan are members of the same family of chemicals, called halogenated aromatic hydrocarbons, which are mostly man-made, poisonous and, by now, widespread throughout the environment. Their unique properties—extreme stability, resistance to heat and attack by other chemicals, hostility to the growth of microorganisms—make them useful for many purposes in industry and agriculture. These same properties, however, have been found to cause unique problems to the environment and to the health of animals and people. All too often one of these chemicals is in the news, as some new discovery of accident or environmental contamination arouses concern over health hazards.

Introduction   7

The extent of the hazards is now the subject of research in laboratories and hospitals around the world. The poisonous effects of these substances are difficult to determine because they are mostly nonspecific and appear gradually. In order to explore and analyze the current state of knowledge, a meeting of scientists was held in June, 1978 by The New York Academy of Sciences. That Conference on Health Effects of Halogenated Aromatic Hydrocarbons, part of a Science Week, forms the basis for this book.

Part II
THE HALOGENATED AROMATIC HYDROCARBONS

Halogenated aromatic hydrocarbons are constructed primarily of carbon and hydrogen, as are most biologic chemicals. They contain one or more ring structures, which give many of them distinctive odors—hence the term aromatic. And they contain chlorine or bromine (halogens), which are foreign to most biologic systems and account for their toxic effects.

Dioxin, the chemical of the Séveso explosion and one of the most poisonous of the group, has the structure:

$$\text{[structure of dioxin with Cl substituents, C, H, and O atoms]}$$

where C=carbon; H=hydrogen; O=oxygen; Cl=chlorine.

Michigan's PBBs, or polybrominated biphenyls, are a mixture of chemical entities, with the general formula:

$$\text{[structure of polybrominated biphenyl with X substituents]}$$

where the X may be either H (hydrogen) or Br (bromine). There are as many as 210 individual structures possible.

The PCBs of the Yusho accident are the same as PBBs except that the X may be H (hydrogen) or Cl (chlorine). Again, there are 210 possible combinations. Most commercial PCBs contain a mixture of 50 or 60 structures.

The granddaddy of the halogenated aromatic hydrocarbon family, ecologically speaking, is DDT, or dichlorodiphenyltrichloroethane, which looks like this:

$$\text{DDT structure}$$

DDT was the first of these chemicals to enter the public consciousness as an environmental hazard. Rachel Carson's *Silent Spring*, published in 1962, called attention to the peculiar properties of DDT and other chlorinated hydrocarbons, arousing such concern that their use as pesticides has been severely curtailed world-wide.* DDT itself was banned in the United States in 1972.

To most biologic organisms, halogenated aromatic hydrocarbons are foreign, unnatural substances, metabolized with difficulty or not at all. They are soluble in fats and oils; most are not soluble in water. Thus they tend to accumulate in the fatty tissue of animals and remain there indefinitely, their concentration increasing with the age of the animals. Fish, in particular, take up chemicals present in water in very small amounts and concentrate them thousands-fold in their flesh. Predators that feed on fish then ingest the pollutants and concentrate them further in their own bodies. Animals at the top of the food chain, such as fish-eating birds and mammals, including man, accumulate the largest quantities of the chemicals and are most likely to suffer harmful effects.

* Many developing countries still use DDT for containment of malaria-carrying mosquitos.

Chapter 1
DDT and Other Organochlorine Pesticides

DDT was first synthesized in 1874, but its insecticidal properties were not discovered until 1939. Its usefulness was immediately apparent, with the result that it was soon being used on a large scale. Other organochlorine insecticides followed shortly: benzene hexachloride (lindane), aldrin, dieldrin, endrin, and chlordane were all introduced in the 1940s. Kepone® (chlordecone) and mirex appeared in the 1950s. Strictly speaking, only DDT belongs to the halogenated aromatic hydrocarbon family, since the others are not considered aromatic; they are closely related, however, and have similar properties.

When organochlorine insecticides were first introduced, they seemed ideal: they were toxic to insects at low doses and appeared to be relatively nontoxic to humans and animals. Their stability seemed an advantage—repeated applications were unnecessary, making them cheap to use. But problems began to appear almost immediately. Fish in lakes and streams died suddenly after heavy rains caused runoff of pesticides from treated croplands. Birds were found dead after DDT was applied aerially to control forest insects. Pesticide residues began to appear in meat and in cow's milk.

In the 1960s, a team of scientists, sponsored by the World Health Organization and the International Agency for Research on Cancer, conducted a world-wide survey to determine the extent to which organochlorine insecticides were getting into the tissues of human beings. They found significant amounts of the chemicals in the body fat of people in Australia, Sri Lanka, Thailand, Nigeria, Kenya, Uganda, South Africa, Brazil, Argentina, Spain, and Israel. Earlier studies had found them in India, North America, and all over Europe. *The conclusion: "DDT is a current constituent of the human body."*

A continuing monitoring program by the U.S. Environmental Protection Agency (EPA) confirms that DDT remains a constituent of American bodies. Samples of tissue taken from patients at surgery or from postmortem examinations are tested for pesticides. During the period 1970 to 1975, almost 100 percent of those tissues sampled had detectable amounts of DDT. But there was good news too—the amounts of DDT appear to be decreasing with time. The phasing out and final banning of the pesticide in the United States has had an effect: in 1970, the mean concentration of DDT in a fat sample was 7.88 ppm (parts per million). By 1974 it had fallen to 4.99, where it remained in 1975. Most of the samples also contained residues of aldrin, dieldrin, benzene hexachloride, and metabolites of heptachlor and chlordane. They were present in smaller concentrations (0.1 to 0.4 ppm), and did not appear to be decreasing with time.

Today the use of organochlorine insecticides in the United States has been essentially discontinued. We continue to export them in large quantities, however. Their economic advantages weigh heavily where demand for food is most urgent. Thus they continue to be a threat to the world environment. In the United States, this group of chemicals has been replaced by others such as organophosphorus insecticides, including parathion and malathion, and carbamates. Some of these are highly toxic to man and require great care during application. They are much less stable, however, and degrade rapidly in the environment.

Chapter 2
PCBs

Now that the use of DDT has come under stricter control, concern has shifted to another chemical, or mixture of chemicals—PCBs, or polychlorinated biphenyls. Heavy oils that are chemically inert, heat resistant, nonflammable, and electrically nonconducting, PCBs have many properties valuable to industry. Their most common application is as dielectrics in transformers and capacitors. They have also been used in lubricating and cutting oils, pesticides, heat exchanger fluids, and as plasticizers in paints, copying paper, adhesives, sealants, and plastics.

PCBs have been used since the 1930s, but until recently they were not regarded as a medical or environmental threat, except perhaps, to those who worked with them. Then in 1966, a Swedish investigator, Dr. Sören Jensen, who was measuring organochlorine pesticide concentrations in wildlife, identified contaminants that were interfering with his measurements— they were PCBs.

Once scientists began to look for these substances in the environment, they found them everywhere, especially in industrialized areas. PCBs have been found in salt- and fresh-water fish, birds of North America and Europe, whales, seals, polar bears, European hares and foxes, and North American mink. Residues are found in the body tissues of Europeans, Israelis, Japanese, and Americans, with Germans having the highest amounts, averaging 7 to 10 ppm in 1974. The Environmental

Protection Agency found that over 90 percent of Americans had detectable levels of PCB in 1974, with about 40 percent having greater than 1 ppm.

Sources of environmental pollution

How did PCBs come to be so widespread? Unlike pesticides, which were sprayed indiscriminately around the environment, PCBs have been used mainly—60 percent in 1970 in the United States—in sealed systems: capacitors, transformers, and heat exchangers.

Some of the worst problems have arisen when these containers have come unsealed. The Yusho disaster is one example. A similar accident occurred in a North Carolina factory in 1971, when a leaky heat exchanger contaminated 16,000 tons of chicken feed over a period of three months before the leak was discovered. The Food and Drug Administration and the Department of Agriculture recalled outstanding stocks of the feed and attempted to trace contaminated eggs and chickens, but only 10 percent of the feed was recovered and an estimated 400 to 1,000 pounds of PCBs were lost directly into human food chains. Levels of contamination were low, however, and no human health problems were traced to the incident.

Another type of accident occurred in April 1977, when a warehouse fire in Puerto Rico caused the rupture of two transformers, each containing 2,000 pounds of PCB-containing oil. The oil contaminated 800,000 pounds of bagged tuna meal, which was later used to make animal and fish feeds shipped mostly to Texas and Arkansas. Again, it was several months before the problem was detected, and much of the contaminated feed was never traced.

The second most common use of PCBs—25 percent in 1970—for plasticizers in paints, inks, copying paper, plastics, adhesives, etc., allows the chemical to reach the environment by a variety of routes. A significant percentage of the plasticizer simply vaporizes into the air. More may enter the atmosphere when discarded products are incinerated, since PCBs are stable up to 800°C. Discarded copy paper may be recycled, allowing PCBs to get into other paper products, including

PCBs

food-packaging materials. PCBs found in cows' milk have been traced to paint on the walls of silos in which cattle feed is stored.

Once the hazards of PCBs became known, such uses were halted. After 1970, the sole U.S. manufacturer, Monsanto Chemical Company, restricted sales for use only in closed systems. Other countries also limited PCBs. Japan banned them altogether in 1972. Production was halted in the United States by 1977. But the chemicals may continue to enter the environment for some time to come as materials sold before 1971 are discarded.

Extent of the pollution

Even if no further pollution were to occur, enough PCBs are already dispersed throughout the environment to cause concern for the indefinite future. The cumulative production of PCBs in North America through 1970 (after which production fell off) has been estimated at 500,000 tons. Worldwide production was about twice that. In North America, an estimated 300,000 tons have been disposed of into dumps and landfills and may or may not be leaking into air and waters. About 30,000 tons have been released into the atmosphere and were probably carried back to earth by rain or snow. And about 60,000 tons were released into fresh and coastal waters.

PCBs have been found in the Great Lakes; Escambia Bay, Florida; the Waukegan River in Illinois; the Ohio River; the Housatonic River in Connecticut; the Chesapeake Bay; San Francisco Bay; Puget Sound, Washington; and in New York's Hudson River. Most of these waters have been polluted by discharge of industrial wastes, either directly or indirectly through municipal sewer systems. The problem is painfully illustrated by the case of the Hudson River.

PCBs in the Hudson River

The problem first came to light in 1975 when fish tested by government agencies were found to contain PCBs at levels far higher than the current Federal Drug Administration (FDA) limit of 5 ppm. In the northern section of the river, small-mouth bass averaged 72.6 ppm, largemouth bass 61.7, and yellow perch 134.6. One goldfish had 1836 ppm of PCBs in its flesh, which amounted to 0.2 percent of its weight.

The source of this contamination was quickly traced to two General Electric Company capacitor plants, which had been discharging large volumes of PCBs into the river for more than 25 years. An immediate ban was placed on commercial fishing in the Hudson, and suit was brought against G.E. for polluting the river with a toxic substance. But the blame has to be shared by Federal and state agencies, since both had granted permits allowing G.E. to discharge 30 pounds a day of chlorinated hydrocarbons into the river.

The settlement, reached in 1976, provided that G.E. would pay $3 million toward cleanup plus another $1 million for research. The state would contribute another $3 million. G.E. promised to reduce its discharges, and by 1977 the company had phased out its use of PCBs entirely. But the damage done will cost much more than the allotted $7 million to repair. By now an estimated 600 tons of PCBs have built up in the Hudson River Basin, of which not quite half is in the river bottom, moved downstream by currents and tides, accessible to fish and community water supplies. The other half is stored in dumps and landfills, subject to erosion and leaching by rainfall into the river bed. Losses to the commercial fishing industry may amount to millions of dollars a year. Sport fishing and illegal commercial fishing, which still occur, may pose a public health problem.

A complete cleanup could cost up to an estimated $204 million. The $7 million has been mostly exhausted in assessing the problem and formulating a plan that would require $25 million more in federal funds. If the money is approved, river sludge in the most heavily contaminated areas will be dredged out and buried in encapsulated landfill sites. This should remove about three-quarters of the PCBs. No one knows how long it would be before contamination of fish would fall to acceptable levels. But if the mass of PCBs is not removed, Hudson River fish will continue to be unsafe to eat in the foreseeable future.

Fish

Although Hudson River fish are among the most heavily contaminated in the country, fresh-water fish from other sources also contain high levels of PCBs. In Great Lakes fish, residues higher than the current FDA limit of 5 ppm are common in several species, and there is no sign that levels are declining. An FDA proposal to lower the permissible limit to 2 ppm would eliminate much of the commercial fishing in the Great Lakes. Crabs in the Upper Chesapeake Bay are also heavily contaminated.

According to the FDA, fish have been the main source of PCBs in the American diet since 1972, when regulations dramatically reduced contamination of other foods. Fortunately, levels of PCBs in the average diet are low, according to the FDA, since salt-water fish are relatively uncontaminated, and over 90 percent of the fish eaten in the United States come from salt water.

Chapter 3
Dioxins and Dibenzofurans; Chlorophenols

Although PCBs are now the most prevalent of the halogenated aromatic hydrocarbons in the environment, the most serious biologic effects may be due to other related chemicals, common contaminants of PCBs. In the Yusho case, for example, the PCBs that contaminated the rice oil were in turn found to contain small amounts of chlorinated dibenzofurans. The tiny amounts of these chemicals are suspected of causing some of the most severe symptoms in the Yusho patients. Other common impurities, also highly toxic, are chlorinated dibenzodioxins, one of which—TCDD—was the chemical of the Séveso explosion.

Dioxins and dibenzofurans are formed from other halogenated hydrocarbons in the presence of oxygen from the air, especially at high temperatures. Thus old, used PCBs contain more of these impurities than the freshly synthesized oil, and are more toxic. Workers are likely to be exposed to the toxic contaminants when working with PCBs at elevated temperatures, for example in steel mills where casting waxes contain PCBs, or while welding transformers or capacitors during manufacture or repair. Incineration of PCB-containing trash (for example, plastics, copy paper) is likely to produce dioxins and dibenzofurans, and in fact, these chemicals have been found in fly ash from municipal incinerators. PBBs, used as fire retardants, form brominated dibenzodioxins and furans at high

Dioxins and Dibenzofurans; Chlorophenols

temperatures. Thus, if a fire-proofed building burns, the fumes may be extremely toxic.

Dioxins are also common contaminants of chlorophenols, an important group of halogenated aromatic hydrocarbons which are produced worldwide at the rate of 200,000 tons a year. Chlorophenols are used as herbicides, fungicides, and wood preservatives and are thus often burned, with the consequent release of dioxins in the smoke. In addition, chlorophenols are used to make the disinfectant hexachlorophene and the herbicides 2,4,5-T and 2,4-D.

Dioxin exposure through industrial accidents

Trichlorophenol for conversion to hexachlorophene was being synthesized in Séveso when the explosion occurred. This reaction has a tendency to "run away" if not kept under control, forming increasing quantities of dioxin as the reaction mixture gets hotter and hotter until it eventually explodes. Such accidents had occurred previously in other factories, and many Séveso victims are bitter because better safety precautions had not been taken. In view of the medical problems suffered by factory workers exposed in previous accidents, Séveso residents should have been evacuated more promptly to minimize dioxin exposure. But officials of the Italian government believe that they were not fully informed of the hazards. The Givaudan Corporation, owner of the plant, did not at first make known the presence of dioxin in the escaped gases. And companies in which previous accidents had occurred were reluctant to divulge the medical hazards for fear of legal consequences.

It is now known that at least fourteen accidents had occurred in trichlorophenol reactors, five of them in the United States. More than 570 people were injured, many developing the typical skin rash as well as damage to internal organs. A study of workers exposed to dioxins through industrial accidents is now being conducted by the International Agency for Research on Cancer at Lyons, France with the cooperation of some of the companies. The health of these workers will be followed to look for long-term effects of exposure to the chemicals. One ominous sign is an indication of an unusually high rate of gastrointestinal cancer among the workers.

Other sources of dioxin exposure

Dioxins, like PCBs, have entered the American food supply through accidental contamination of animal feeds. One such incident occurred in the late 1950s, when millions of chickens died of so-called chick-edema disease, with symptoms including droopiness, ruffled feathers, difficulty in breathing, and accumulation of fluids. The cause of the disease was traced to contaminated fats added to the feed, and the FDA ruled that animal feed must henceforth be free of "chick-edema factor." Several years elapsed, however, before this factor was identified as a dioxin. The fats used in the feed were by-products of the tanning industry, which treated hides with dioxin-contaminated chlorophenol before trimming off the fat. In this incident, and in other instances of chick-edema disease that have occurred since, the toxic contaminants were probably passed along to people who ate eggs and meat from the contaminated poultry.

People and animals have also been exposed to dioxin through the use of contaminated oil for dust control. A famous episode occurred in Missouri in 1971, following the spraying of oil on three riding arenas. Over sixty horses died, along with hundreds of birds and dozens of pet dogs and cats that lived in adjoining stables. Several people became ill, including a six-year-old child who frequently played on the arena floor. She was hospitalized for hemorrhaging of the bladder and inflammation of the kidneys. The horses that survived had serious breeding problems, including spontaneous abortions, stillbirths, and deformed foals. Only one foal survived out of 41 breedings that year. The poison was traced to the waste products from a trichlorophenol factory, which had been collected by a salvage oil company and mixed with discarded motor oil and lubricants from service stations for use in dust-control spraying.

Agent Orange in Viet Nam

Thousands of people may have been exposed to dioxin during the Vietnamese war when TCDD-contaminated Agent Orange was used as a defoliant by the United States. It is estimated that over eleven million gallons of the herbicide were sprayed over Vietnamese forests and rice fields during the late 1960s, and that this included perhaps 220 pounds of dioxin. Effects on the Vietnamese population have been hard to document, given the disorganized circumstances caused by the war. Increased birth defects and liver cancer have been reported. Since a longer latent period is usually required for chemically caused cancer to appear, however, and since liver cancer is common in Southeast Asia anyway, many scientists doubt that cancers reported in Viet Nam were caused by dioxin.

Some American veterans of the war, exposed to the chemical during spraying operations, now claim to be suffering from dioxin poisoning. The problem came to light in 1977, when a Veterans' Administration claim worker in Chicago began to notice a peculiar pattern of symptoms among men she interviewed: chronic skin rashes, arthritis-like soreness, stomach and liver problems, numbness in the limbs, extreme fatigue. The veterans' rights group, Citizen Soldier, claims that many more veterans may be affected by the exposure but remain unaware of the cause of their troubles.

A survey by Citizen Soldier in collaboration with scientists from the American Health Foundation in New York City has elicited about 1200 responses from veterans, most describing the typical symptoms. Some report cases of cancer or birth defects, but further statistical analysis is needed before the scientists can say whether these problems are linked to dioxin exposure.

Herbicide use in the United States

Use of organochlorine herbicides has not been confined to Viet Nam. The two components of Agent Orange—2,4,5-T (2,4,5-trichlorophenoxyacetic acid) and 2,4-D (2,4-dichlorophenoxyacetic acid)—have also been used extensively in the United States, sprayed on forests, pastures, rangeland, and along roadways and power lines to kill undesirable species of plants. 2,4,5-T and 2,4-D differ from most other halogenated aromatic hydrocarbons in being soluble in water and are thus excreted rapidly from the body. They are also degraded more rapidly in the environment and thus escape the major drawbacks of their chemical relatives.

In 1970, however, highly publicized reports that 2,4,5-T caused birth defects in laboratory animals led the government to restrict its use to prevent contamination of food and water. The evidence against 2,4,5-T has been confused by the fact that in the past, most commercial preparations were highly contaminated with TCDD. Preparations now on the market generally contain less than 0.1 ppm TCDD. 2,4,5-T was in the news again in early 1979, however, with evidence that linked its use to high rates of miscarriage among women living near forests sprayed with the herbicide. The Environmental Protection Agency immediately banned most applications of 2,4,5-T as well as the closely related herbicide Silvex. Further controversy is likely before a decision is reached on whether to make the ban permanent.

2,4-D has not been found to contain TCDD, although it may sometimes be contaminated with other dioxins. In addition to nonagricultural uses shared with 2,4,5-T, 2,4-D is used to kill weeds infesting crops such as wheat, corn, and sugar cane, and to control the ripening of some fruits. The safety of 2,4-D has also been questioned, and the herbicide is on the Environmental Protection Agency's list for future consideration.

Chapter 4
PBBs

PBBs have been identified as an environmental problem primarily in Michigan through the animal feed mix-up. Their production, distribution, and use—as fire retardants in television casings, light housings, typewriters, and office machines—is much more limited than that of PCBs. Since they are solid at room temperature and do not migrate from the thermoplastics into which they are incorporated, they are less likely to enter the environment.

Nevertheless, recent studies of areas surrounding PBB industries in Michigan, New Jersey, and Arkansas found PBBs and other brominated hydrocarbons in air, soil, and water samples. In Arkansas, brominated chemicals were even found in hair collected from a barber shop, indicating that the people had somehow been exposed.

In Michigan, the 1973 accident resulted in contamination of soils through disposal of PBB-tainted manure, milk, feed, carcasses, and dust cleanings from buildings. The chemicals do not break down in the soil, their solubility in water is low, and they are not taken up by plants; so PBBs will probably remain as a permanent component of the soil.

A greater problem stems from the Michigan Chemical Corporation's disposal practices during the years 1970 through 1974, when it was the country's major manufacturer of PBBs. The substances were discharged with the factory's wastewater into the nearby Pine River, where fish have concentrated the chemicals to dangerous levels in their tissues. Furthermore, the landfill site where the company disposed of 269,000 pounds of waste, containing 60 to 70 percent PBBs, was a poor choice geologically. Studies have shown the chemicals to be leaching out into surrounding groundwater, perhaps to threaten drinking water supplies in the future.

Chapter 5
Other Pollutants

Undoubtedly, other halogenated aromatic hydrocarbons are polluting our environment, our water, and our food supplies. We may not yet be aware of them, either because they are less widespread, because no major disaster has been attributed to them, or perhaps merely because we do not yet have the means to identify them.

Industrial waste disposal

Industrial waste disposal is an enormous source of pollutants. According to the Environmental Protection Agency (EPA), 30-40 million tons of hazardous waste are produced in the United States each year, and 80 to 90 percent of them are disposed of in unsatisfactory ways. PCBs and PBBs are only some of the chemicals that have entered public waterways in factory effluents. Other halogenated hydrocarbons have been found in the wastewaters of industries that manufacture petrochemicals, plastics, textiles, automobiles, electronic components, and pharmaceuticals. Although Federal law now limits discharge of effluents directly into waterways, many industries discharge them into municipal treatment facilities, which are not equipped to handle chemical pollutants and in turn discharge them into waterways.

Other disposal methods are scarcely more satisfactory. Residents of Niagara Falls, New York are now suffering the consequences of Hooker Chemical and Plastics Corporation's 20-year practice of burying wastes in metal drums in an abandoned canal. In 1953, disposal was halted and the site was filled in and developed into a residential area. But the drums began to leak and, by 1978, backyards and basements were oozing chemicals, 82 in all, 11 of them suspected of causing cancer. Residents suffered burns and possible increased rates of birth defects, miscarriages, and liver damage. Eventually the area was declared a Federal disaster area and evacuated. To make matters worse, the chemicals may be seeping into the Niagara River, a feeder for Lake Ontario, which provides drinking water for millions of Americans and Canadians. This site is only one of the 638 chemical dumping sites in the United States that the EPA has identified as "imminent hazards to public health."

Water pollution

In addition to industrial wastes as a major environmental problem are the "nonpoint" sources, which account for about half of the nation's water pollution. This pollution consists primarily of runoff water and includes wastes from agriculture, lumbering, mining, and construction. Urban runoff may contain pesticides and fertilizers from construction sites, lawns and gardens, oily residue from automobiles, street litter, air pollution particulates, and sewer overflow.

Point and nonpoint sources of chemicals combine to pollute drinking water sources, especially rivers. Most treatment systems, predominantly aimed at eliminating microbial contamination, are not able to remove chemicals from water. In fact, treatment with chlorine for disinfection may make chemical pollution worse, since the chlorine may react with organic chemicals already present to produce toxic chlorinated compounds. Studies by the EPA and others have identified more than ninety chlorinated organic chemicals in U.S. drinking water, along with many other impurities. Noxious chemicals are not found in every water sample, of course, and the concentrations are usually very low. But the long-term health effects of this pollution are still unknown.

Part III
HEALTH EFFECTS

Most of our knowledge of the health effects of halogenated aromatic hydrocarbons has come from observing people exposed to large quantities of the chemicals, either occupationally or through some sort of accident. The relevance of this information to the majority of the population that receives much lower exposures chronically, perhaps over a whole lifetime, can only be guessed.

Even where people were exposed to high doses, it is often difficult to relate specific symptoms to specific chemicals. In most instances, exposure has been not to a pure substance, but to a mixture. Although many of the chemicals in this family cause similar symptoms, they differ greatly in their toxicity. Dioxin (TCDD), for example, is so poisonous that 22 millionths of a gram per kilogram of body weight will kill half the male rats receiving that dose. PCBs are almost a million times less toxic: 4 to 10 grams per kilogram are required for the same effect. PCBs are a mixture of up to 210 possible chemical entities, differing in the number and location of chlorine atoms on the basic carbon frame; both factors influence their biologic effect. Moreover, commercial PCBs are contaminated with dioxins and dibenzofurans, which may be responsible for some of the symptoms attributed to the PCBs.

Although harmful effects of chemical exposure may sometimes be painfully obvious, detailed scientific methods are often necessary to find the specific biologic effects of chemicals. Epidemiologic techniques may be required to identify the cause of an unexplained outbreak of symptoms, as in the Yusho incident, where patients were extensively questioned on all aspects of their lives before the cause of the problem was finally identified as a certain brand of cooking oil. Sometimes the symptoms themselves are vague or occur widely in the general population, such as "tiredness," "headache," "insomnia." In these cases it is difficult to sort out how much of a problem there actually is. This occurred in Michigan, where investigators tended to blame complaints on stress or psychologic factors until it was discovered that virtually everyone in Michigan had been exposed to PBBs. Only after a group of Wisconsin residents was chosen as the control group—the "normals"—could harmful effects of PBBs be identified.

Experiments on animals are another way of relating biologic effects to specific suspect chemicals. Laboratory experiments are also able to look for underlying disturbances of biologic function, which may cause only vague symptoms or none at all. Different species sometimes show different responses to chemicals, however, and the results of animal studies are not necessarily applicable to humans. For example, halogenated aromatic hydrocarbons cause liver damage in most animals. But the guinea pig, perhaps the most sensitive species to these chemicals, may die of the poisoning with an intact liver.

Chapter 6
General Symptoms

The most common and conspicuous symptom of exposure to halogenated aromatic hydrocarbons is chloracne, a form of acne whose association with chlorinated chemicals has been recognized as an occupational hazard since 1899. In the 1930s, when chlorinated hydrocarbons gained popularity as electrical insulating materials, workers began suffering from "cable rash," "Aroclor acne" (Aroclor is a trade name for PCBs), "Halowax acne" (Halowax is a trade name for chlorinated naphthalenes) and "electrician's rash." Occasional workers suffered more severe symptoms, including liver damage and neurologic disorders, and a few deaths were attributed to Halowaxes, Aroclors, and organochlorine pesticides. Nonetheless, these chemicals were generally regarded as not too toxic in low doses.

PCBs

The Yusho incident, coupled with the discovery that PCBs were widespread in the environment, aroused new interest and concern about their adverse health effects. Like occupationally exposed workers, the Japanese patients had chloracne, especially of the face and genitals. They also developed black pigmentation of the face, eyelids, lips, and gums. Nails on fingers and toes were discolored and often deformed, and swelling and discharge from the eyes were common. Infants born to Yusho mothers showed a dark brownish pigmentation over their entire bodies. This discoloration disappeared gradually over a period of months.

Yusho patients also complained of more general symptoms such as loss of appetite, nausea, vomiting, weakness and numbness of extremities, but these were at first regarded as complications of minor importance. Over the ten years since the accident, however, although the skin symptoms greatly improved, the more general symptoms persisted and grew worse. Patients still suffer from dullness, headache, indefinite stomach-ache, numbness and pain of extremities, swelling and pain in the joints, coughing, and bronchitis-like symptoms.

Stimulated by the Yusho findings, epidemiologists have begun intensive studies of workers exposed to PCBs on the job. A research team from the Mt. Sinai School of Medicine in New York City recently conducted an investigation of workers exposed for an average of fifteen years in a capacitor manufacturing plant. Forty percent of them were found to have abnormalities of the skin, including acne and pigmentation. Thirty percent complained of headaches. Other symptoms in common with Yusho victims included discharge from eyes and swelling of eyelids, musculoskeletal symptoms, and gastrointestinal symptoms. Nine to 15 percent had abnormalities that showed up in lung function tests; many of these also had respiratory symptoms, such as cough and tightness in the chest.

Hexachlorobenzene

Other chlorinated hydrocarbons cause many of the same symptoms as PCBs. In one of the worst disasters caused by an environmental chemical, more than 3,000 residents of southeastern Turkey were poisoned in the late 1950s by hexachlorobenzene, which had been used as a fungicide on seed grain distributed by the government for planting purposes. It was a time of scarcity, and the people had used the seed grain for food. Ten percent of the victims died. Infants and children were especially vulnerable, and in many villages in the years 1955 to 1960, no children remained between the ages of 2 and 5.

Symptoms in this Turkish population were similar to other

General Symptoms

epidemics of halogenated hydrocarbon poisoning, but more severe. Most conspicuous was severe blistering and unusual sensitivity of the skin to light and minor mechanical irritation. The outer layer of skin rubbed off easily; ulcers formed, which healed poorly and became infected; scarring was common. Children were called "monkey children" by the peasants because of their dark pigmentation and a fine layer of dark hair that appeared on the face, trunk, and extremities.

Many patients developed arthritic deformities in the fingers and some had neurologic symptoms making them unable to use eating utensils, rise from a squat, or climb stairs. Loss of appetite, weight loss, and wasting of skeletal muscles were common. Strikingly, the urine turned a port-wine color, symptomatic of liver disease, and the liver itself was enlarged. Although the most severe symptoms abated among surviving patients when the poison was removed from their diet, many victims are still affected twenty years later.

Dioxin

In Séveso, 447 people suffered from chemical burns within the first days after the accident. Chloracne takes longer to develop and appeared 4 to 6 weeks later in about 10 percent of the 733 persons most heavily exposed to the dioxin. Most of these cases were among children and teenagers. Screening of school children in surrounding, less contaminated areas, led to the identification of about 70 additional cases, most very mild. Many of the Séveso residents experienced nausea, vomiting and headache in the days following the accident, but investigators claim not to have found any persistent symptoms that could be attributed to the contamination. Screening of children found enlargement of the liver in about 10 percent and visual problems in about 8 percent.

PBBs

In Michigan, farmers whose herds had been devastated by PBB poisoning began complaining of vague symptoms: headaches, rashes, numbness, loss of appetite. After the first survey by the Michigan Department of Health found no significant differences between 189 adults known to have been exposed and a control group of other Michigan residents, a more elaborate study was begun, in cooperation with the Center for Disease Control in Atlanta, Georgia. A total of 4,545 persons were divided into four groups: persons living on quarantined farms, persons who received food directly from these farms, PBB workers and their families, and a miscellaneous group believed to have received only low-level exposure. Tests for PBBs in the blood were done, and the first three groups did indeed have much higher levels. But although rashes, fatigue, and joint pain were found to be common in the three exposed groups, many of those with low PBB levels in their blood also had the same symptoms.

Another investigation by the Mt. Sinai medical team has clarified the situation somewhat: 933 farmers and Michigan residents likely to have eaten contaminated farm products were compared with a group of 228 farmers from Wisconsin, where PBB contamination had not occurred. With this approach, differences became obvious. Neurologic symptoms were reported much more frequently in Michigan than in Wisconsin. Extreme fatigue was a common complaint—many persons reported sleeping 14, 16, or 18 hours a day. Arthritis-like symptoms were reported in young Michigan men. Skin problems were also more common in Michigan. Blood tests found abnormal enzyme levels in many of the Michigan residents, especially those who complained of symptoms.

Two recent studies of randomly selected Michigan residents, one by the Mt. Sinai team and one by the State Health Department, have found detectable levels of PBBs in the blood of about 90 percent of the population of the state, and in the fat tissue of almost 100 percent. PBBs were not found in the blood of a group of people who had moved to Michigan only recently, which suggests that the state's food supply is no longer contaminated. Scientists are still surprised by the lack of correlation between the amount of PBBs in the blood and the severity of the symptoms.

General Symptoms

Animals

Many of the complaints by victims of halogenated aromatic hydrocarbon poisoning are vague and easily dismissed as psychosomatic. But animals exposed to the chemicals exhibit a variety of similar symptoms. Monkeys fed large amounts of halogenated aromatic hydrocarbons respond much like human patients. They develop chloracne on the face and lips, swelling of the eyelids, and discharge from the eyes. Skin on the rest of the body becomes dry and scaly, and finger- and toe-nails may fall off. The animals lose their appetite, lose weight, become progressively weaker, and eventually die. Rats and guinea pigs show few overt symptoms, merely a gradual weight loss and wasting away.

Cattle have occasionally been poisoned through accidental addition of halogenated aromatic hydrocarbons to feed, or even by being housed in structures painted with wood preservatives. They develop "X-disease" with symptoms that include diarrhea, chronic cough, poor appetite, discharge from eyes and nose, and hardening of the skin with cracks and wart-like growths. These were also the symptoms of the Michigan cows poisoned by PBBs.

Rabbits and hairless mice are the only animals other than man and monkeys that get chloracne. If chemicals are painted on the inside surface of a rabbit's ear, or on the skin of a hairless mouse, the response is much like that of human skin. Scientists use this method to sort out the different effects of individual chemicals. Trichlorophenol, for example, has long been known to cause chloracne among workers. Experiments with rabbit ears have shown, however, that pure trichlorophenol does not give a response, although the technical grade chemical does. It is now believed that dioxins and dibenzofurans, present as impurities, are the cause of the chloracne. Both give a response with the rabbit-ear test.

Such methods have shown that, in general, more highly chlorinated (or brominated) chemicals are more harmful. But the arrangement of the chlorine or bromine atoms is also important. The rabbit-ear test has shown, for example, that chlornaphthalenes containing 5 or 6 chlorine atoms cause chloracne, whereas molecules containing 1,2,3,4,7, or 8 chlorines do not. Such observations are difficult to explain.

Chapter 7

The Search for Underlying Metabolic Disturbances

Scientists still do not understand why animals die from consumption of halogenated aromatic hydrocarbons. Postmortem examinations of experimental animals find that effects on various organs differ among species, and none of the damage seems severe enough to cause death. The only consistent findings in all species are almost complete loss of body fat, shriveling of the thymus gland, and enlargement of the liver.

Damage to blood vessels causes collection of fluid in the abdominal cavity and heart area of chickens and some other animals. Blood vessels in the stomach and small intestine may also be injured, ulcerate, and hemorrhage. In monkeys, changes in the stomach cells similar to early stages of cancer are common. Monkeys and guinea pigs show signs of kidney damage. The testes degenerate in many species. In all animals, the bone marrow loses cells to some degree, often leading to anemia.

Liver function

In the search for an underlying metabolic disturbance that would explain the generalized effects of halogenated aromatic hydrocarbons, much attention has focused on the liver. Liver function is affected by these chemicals in all species, including the guinea pig and man, although not all the changes are signs of damage. Effects include enlargement of the organ as a whole and an increase in certain components of cells responsible for synthesizing enzymes. Frequently, enzyme levels in the blood are also altered, reflecting the liver changes. Such disturbances were observed among some of the Séveso victims as well as in the Michigan population.

Among Yusho patients, abnormally high levels of triglycerides were found in the blood, suggesting that PCBs had disturbed fat metabolism in the liver. High cholesterol levels have been measured in some occupationally exposed persons. Definite signs of liver damage have been observed, for example, among workers exposed to Halowaxes in the 1930s and 1940s, when cases of jaundice were reported. Decreased tolerance to alcohol has been reported by PCB-exposed workers and by Yusho patients.

A disease called porphyria cutanea tarda, occurring among chlorinated hydrocarbon workers, is sufficiently well-known to be found in medical dictionaries. Its symptoms include the familiar acne, sensitivity of the skin to light and mechanical irritation, and "port wine urine," caused by the presence of porphyrins in the urine. The porphyrins, which are intermediates in the synthesis of the blood pigment hemoglobin, are produced in excess by the liver when its function is altered by the chemical stimulants. Full-blown cases of porphyria cutanea tarda occurred among the Turkish victims of hexachlorobenzene poisoning. Urinalysis of Michigan residents exposed to PBBs and Séveso residents exposed to dioxin indicates that some of these individuals had a mild form of the disease 2 to 4 years after the exposure.

Enzyme induction

Since the liver is the bodily organ primarily responsible for metabolizing foreign substances, some of the changes observed are beneficial responses to the threat posed by the foreign chemicals. Liver enlargement is merely a sign that processing facilities are being expanded to handle the increased load. Specific enzymes, which are protein molecules, are synthesized by liver cells for the purpose of guiding specific chemical reactions, intended to convert the foreign material to a form that can be excreted more easily. This enzyme-induction process is part of a finely tuned mechanism whereby the cell can respond

The Search for Underlying Metabolic Disturbances

appropriately to different situations. A certain nutrient or foreign chemical stimulates the cell's protein synthesizing machinery to produce the enzymes necessary for the cell to metabolize that nutrient or chemical. The mechanism is more or less the same in most experimental animals and man.

Although this process works well for many foreign chemicals, flaws in the system show up especially with many halogenated aromatic hydrocarbons. Enzymes are induced in groups, depending on the three-dimensional structure of the stimulative chemical. Only two major enzyme groups are available to deal with foreign chemicals and, for the most part, neither of them is very efficient at metabolizing the halogenated hydrocarbons. This accounts for the long-term persistence of the chemicals in the body. As long as the chemical is present, some of the enzymes will also remain at high levels. They may therefore participate in the metabolism of other chemicals, which would normally be handled differently by the body. For example, PCBs induce the enzyme system that also metabolizes phenobarbital in rats. Thus the length of time a rat will sleep after it is given a certain dose of phenobarbital is decreased if it has previously been fed PCBs.

The enzymes also metabolize some hormones. Halogenated hydrocarbon exposure leads to reduced blood levels of sex hormones in many species. This may affect sexual characteristics, causing, for example, the shrinking of seminal vesicles in mice and a reduction in weight of testes, comb, and wattles of roosters. The lower levels of sex hormones may also interfere with reproduction. Birds are especially sensitive to this effect, since normal eggshell production depends on blood estrogen levels. Even small decreases in the hormone leads to thinning of eggshells and thus to easy breakage.

Thyroid hormone is also metabolized at an increased rate by animals fed PCBs, as is cortisone. Vitamin A levels are generally low in animals exposed to halogenated hydrocarbons, perhaps as a result of increased metabolism of the vitamin.

Scientists are still not certain whether the enzyme response to halogenated hydrocarbon exposure is the most significant effect. But they suspect that some such mechanism may explain why the chemicals can cause death. Somehow the chemicals must interfere with fundamental processes in the cells, leading to a general malfunction of those cells. When this happens, the animal appears to starve to death.

Chapter 8
Neurologic Effects

The most common complaints among people who have been accidentally poisoned by halogenated hydrocarbons are neurologic: headache, numbness or pain in extremities, muscle weakness, disturbances in vision. At least half of the Yusho patients complained of such symptoms, although doctors at first discounted their importance. In Michigan, according to the Mt. Sinai investigators, the earliest and most prominent symptoms reported were neurologic.

Concrete evidence exists that halogenated hydrocarbons can damage nerves, even the brain. DDT and other chlorinated insecticides function as nerve poisons. Although the mammalian nervous system seems to be much less sensitive to the chemicals than that of insects, neurologic symptoms are still the most important sign of pesticide poisoning. Animals poisoned by DDT become nervous and hyperexcitable, then develop muscular weakness, tremors and eventually convulsions, paralysis, and death. Similar symptoms have occurred in people poisoned by DDT, aldrin, dieldrin, endrin, telodrin, and the herbicide 2,4-D.

The persistence of the neurologic symptoms among Yusho patients ten years later forced investigators to take them seriously. Minor nerve damage is difficult to detect in human beings, but tests did show a slowing of the velocity of nerve conduction in some of the patients. Similar tests on Séveso residents exposed to dioxin also showed slowing of nerve conduction velocity. Séveso victims also complained of numbness and pain in extremities, muscle weakness, and disturbances of vision.

PBBs

In Michigan, two recent investigations into the neurologic effects of PBB exposure obtained conflicting results. Physicians at the University of Michigan Medical Center in Ann Arbor and Henry Ford Hospital in Detroit studied 46 patients with incapacitating health complaints blamed on the PBB accident. They concluded that 67 percent of the patients suffered from depression, probably brought on by the stress of being victims. They also found abnormal nerve conduction in 41 percent of the patients. A variety of other psychologic and neurologic tests gave normal results.

The Mt. Sinai team claimed to find many more significant effects by comparing Michigan patients with unexposed Wisconsin residents. Both groups were asked to fill out vast questionnaires describing their physical and psychologic health over the past several years. They were then subjected to a battery of performance tests designed to assess brain function. Michigan residents described many more neurologic symptoms than comparable Wisconsin residents, and they did less well on the brain function tests. The symptoms included tiredness, depression, headaches, difficulties in balance, blurred vision, and muscle weakness. People who suffered the most marked symptoms were the ones who showed diminished performance on the tests. The symptoms could not be blamed on stress, say the investigators, since consumers of products from quarantined farms suffered as much as the farmers themselves, who had had to destroy their herds and face financial disaster.

Experiments done at the National Institute of Environmental Health Sciences demonstrated that PBBs can cause comparable symptoms in animals. Experimenters found that rats fed Firemaster were less active than normal and developed muscle weakness in the legs. This is analogous to complaints by Michigan residents of muscle weakness, apathy and lethargy, said the investigators.

Unsuspected damage

The fact that neurologic symptoms are ambiguous and difficult to substantiate, even among people known to have been exposed to chemicals and who complain of feeling sick, means that harm caused by lower exposures would almost certainly go unnoticed. Dr. Herbert Schaumburg of the Albert Einstein College of Medicine in New York City has raised the possibility that neurotoxic chemicals in the environment are causing health problems of which we are unaware. Long-term exposure to the chemicals can damage nerves of experimental animals even when the concentrations are below what we might accept in an environmental pollutant, he says. The effects of environmental chemicals on the nervous system are subtle and do not become apparent for years. If the brain is being damaged, there is almost no way of knowing it—vision and memory fail, but that happens anyway as we get older. According to Dr. Schaumburg, environmental chemicals may be enhancing the aging process.

Another possibility of unsuspected damage by halogenated hydrocarbons was raised at the Science Week Conference by Dr. Samuel Chou of the West Virginia University Medical Center in Morgantown. He and his co-workers have found a peculiar behavior pattern in the offspring of mice fed tetrachlorobiphenyl, a PCB, during pregnancy. The dose of chemical was not high enough to cause symptoms in the mother mice, although it did cause the death of many of the mouse pups. Over half of the surviving pups developed a "spinning syndrome," chasing their tails at rates of 40 to 150 turns per minute. They carried on this activity for most of the day throughout their lifetimes. Dr. Chou suggested that this behavior may be analogous to hyperactivity in children, which might similarly be caused by prenatal exposure to low levels of PCBs or other chemicals.

Hexachlorophene

A familiar member of the halogenated aromatic hydrocarbon family, hexachlorophene, is now believed capable of causing severe brain damage. During the 1950s and 1960s, hexachlorophene was used routinely in soaps, especially in hospitals, because of its antibacterial qualities. Although the chemical was known to be poisonous when ingested, it was not known to be absorbed through the skin. Then several incidents occurred that alerted doctors to the possible harm caused by routine washing.

Because of a proposed change in regulations to allow hexachlorophene as a fungicide on fruit and other foods, the FDA undertook several studies on its toxicity. Rats were fed the chemical and developed typical neurologic symptoms of muscle weakness and, eventually, paralysis. Postmortem examination revealed an abnormality in the brain and spinal cord called status spongiosus, or spongy state, because of its appearance under a microscope. But the most alarming results came from a study, done in 1971, in which infant monkeys were washed daily for ninety days with a hexachlorophene soap. When the monkeys were autopsied at the end of the experiment, they all were found to have status spongiosus in their brains.

Evidence mounted that the chemical could be absorbed through the skin under conditions routine in hospital nurseries. Studies found hexachlorophene in the blood of infants, especially premature infants, who had been washed with the soap. Autopsies on premature infants who had died from a variety of causes found status spongiosus in the brains of those who had had the greatest exposure to hexachlorophene. The condition was never seen in infants who had not been exposed to hexachlorophene, although it has also been produced in animals by other chemicals.

The spongy appearance giving rise to the term status spongiosus is caused by the accumulation of fluids in certain parts of the nervous tissue. Whether the condition causes any permanent damage is not known. No symptoms have been noted in normal infants bathed routinely with hexachlorophene soaps. In experimental animals many of the symptoms regress after exposure to the chemical is halted.

Hexachlorophene baths have proved fatal, however, when skin damage permitted increased absorption of the chemical.

Neurologic Effects

In 1973, four instances were reported in the United States in which children with impaired skin died as a result of being bathed for three or more days with hexachlorophene soap. Two of the children had burns, though not serious ones, and two others were infants born with congenital ichthyosis, or "fish skin disease." All four children were found to have status spongiosus in the brain.

A major episode of hexachlorophene poisoning occurred in France in 1972, after the chemical was accidentally added to baby powder at double the normal concentration. Forty-one infants and young children apparently died of the poison. Absorption of the chemical was probably enhanced because the powder was used on skin inflamed with diaper rash and then covered with a diaper.

Needless to say, use of hexachlorophene is now strictly curtailed in the United States.

Chapter 9
Immunologic Effects

Another of the halogenated aromatic hydrocarbons' subtle effects may be to increase susceptibility to infection by damaging the immune system, the body's natural defense against disease. Shriveling of the thymus gland, an organ of the immune system, is one of the most consistent signs of poisoning by these chemicals. Other immunologically important tissues such as the spleen and lymph nodes are also commonly affected.

Types of immune response

Little concrete evidence is available on what chemicals do to immunologic function, partly because the immune system itself was poorly understood until quite recently. Now we know that there are several immunologic mechanisms, some nonspecific, or general, mechanisms that kill or prevent multiplication of microbes and other parasites. Specific or acquired immunity, of which there are two types, allows the animal to "remember" a foreign substance and react more strongly against that substance on repeated exposure.

One type of specific immunity is called humoral immunity and includes the classical antigen-antibody response. The antigen, or foreign substance, stimulates so-called B-lymphocytes to produce antibody, a protein molecule specifically able to combine with and inactivate the antigen. Nonspecific mechanisms then take over and destroy the complex.

Immunologic Effects

The other type of specific immunity, cell-mediated immunity, is the one that seems to be most susceptible to damage by halogenated hydrocarbons. Cell-mediated immunity works through T-cells, thymus-derived lymphocytes, which react directly with antigens. T-cells are, like B-cells, produced in the bone marrow, but the former are in some way activated by the thymus gland.

Effects of chemicals

Dioxin (TCDD) has been shown to be an especially potent suppressor of the immune system in a variety of tests. It causes atrophy of the thymus in every species studied; it interferes with rejection of skin grafts in rats and mice; it decreases the response to a tuberculin skin test in guinea pigs; and very small doses drastically increase the susceptibility of mice to infection with *Salmonella* bacteria.

Some experimenters have also found that PCBs and PBBs interfere with the immune response, primarily through cell-mediated immunity. On the other hand, hexachlorobenzene stimulates humoral immunity in rats, as shown by increased amounts of antibodies in their blood after exposure to the chemical. In mice, however, hexachlorobenzene depresses both cell-mediated and humoral immunity.

There are other factors that further confuse the issue. Immune response is sensitive to hormones, including corticosteroids, thyroid hormones, and estrogens. Because chemicals may alter blood levels of these hormones, they may indirectly affect the immune system by this means. Furthermore, symptoms of halogenated hydrocarbon poisoning are similar to those of malnutrition, which itself can interfere with the immune response.

Young animals are the most sensitive to immunologic impairment by chemicals. This may correlate with the fact that the thymus gland becomes less important and tends to atrophy as an animal gets older—an adult can function quite well without a thymus.

Observations in human beings

There is very little evidence on the effect of halogenated hydrocarbons on the immune system of human beings. Yusho patients were reported to have some changes in antibody levels in the blood two years after exposure; however, the levels subsequently returned to normal. The patients also suffered a high rate of chronic bronchial infection, possibly but not necessarily because of immunologic impairment. Abnormalities of lymphocyte function were found in the Michigan farmers exposed to PBBs, according to the Mt. Sinai investigators.

In Séveso, 5 to 10 percent of the victims of dioxin poisoning were reported to show a transient reduction in the number of lymphocytes in their blood, but other immunologic tests were normal. Investigators noted an increase in reports of infectious childhood diseases, but attributed this to greater attentiveness of the doctors in reporting these symptoms.

Chapter 10
Carcinogenicity

What worries people most about environmental chemicals is the possibility that they may cause cancer. But the relationship between long-term, low-level exposure and the development of cancer is extremely difficult to pin down. People who have been exposed accidentally or occupationally to carcinogens often escape all symptoms at the time of their exposure and yet have a greatly increased risk of developing cancer twenty or thirty years later. The small amounts of the chemicals that we all carry in our bodies for years may cause an increased rate of cancer in the general population, although no one knows how great the risk might be. The interplay of factors in the development of cancer is complex, and the evidence available on the carcinogenicity of halogenated hydrocarbons is scarce. Nonetheless, enough is known to cause concern.

Animal studies

Several of the halogenated hydrocarbons have been found to cause liver tumors when fed to rodents: DDT, dieldrin, mirex, Kepone, hexachlorobenzene, PCBs, PBBs, TCDD. TCDD has also produced tumors of the lungs and mouth in rats, and tumors of the skin when painted on the backs of mice.

The carcinogenicity of these chemicals may be related to their other effects on the liver. Most of the animals that develop liver tumors also have other forms of liver damage, and there is some controversy over what types of damage may be precancerous. For example, porphyria, which occurs in humans exposed to TCDD, hexachlorobenzene, and PBBs has been found in animals with liver cancer induced by TCDD. Whether the porphyria is connected with the cancer, or whether the two phenomena are independent effects of the chemicals, is not known.

Liver enzymes play an important role in the induction of cancer. These enzymes may inactivate harmful chemicals, or they may transform chemicals that are not themselves carcinogenic into by-products that can cause cancer. Most of the halogenated aromatic hydrocarbons are believed to need activation before they become carcinogenic.

Because of the enzymes' role, the response to carcinogenic chemicals is often complex. Treating an animal with one chemical may induce a certain enzyme system, which may change the animal's susceptibility to a second chemical. The timing of exposure may be very important. For example, PCBs fed to rats before administration of a known carcinogen may inhibit the development of liver tumors, whereas if the carcinogen is given first and the animal is then fed PCBs, the likelihood of cancer is increased. Presumably, this type of interaction occurs in the real world, where we are all constantly exposed to small amounts of many carcinogens. It is hard to predict the overall effects of these complex exposures.

Indirect effects

Some types of cancer are strongly dependent on hormones, and the tendency of halogenated hydrocarbons to alter hormone levels may thus indirectly alter the risk of cancer. A study by scientists from the Dow Chemical Company that showed TCDD to cause tumors of the liver and lung in rats, also showed a decrease in spontaneous tumors of the pituitary gland, uterus, mammary gland, pancreas, and adrenal gland, perhaps due to hormonal effects. Chemicals may have different effects in animals of different sexes. For example, in rats, only females develop porphyria and liver tumors in response to TCDD;

Carcinogenicity

males develop neither. Hormones could be the factor responsible for the difference.

Halogenated hydrocarbons may also indirectly affect carcinogenesis through their effects on the immune system. A defective immune system increases the likelihood of developing cancer. This is true of people born with immune deficiencies and also of patients treated with immunosuppressive drugs for organ transplants or for other medical reasons. Although the reason for this effect is not understood, one theory holds that transformation of normal cells into cancer cells occurs frequently in the body, but that the immune system is able to recognize the transformed cell immediately and destroy it. An immune system damaged by exposure to chemicals, for example, would not respond as efficiently to the transformed cell, which might then develop into a tumor.

Cancer in human beings

Evidence on the development of cancer in persons exposed to halogenated aromatic hydrocarbons is limited and inconclusive. Most of the familiar incidents have been too recent for carcinogenic effects to appear. Of approximately 1200 Yusho patients exposed ten years ago, 51 are known to have died, and the cause of death is known for 31 of them. Of these 31, 11 died of cancer, including cancer of the lung, liver, stomach, breast, and malignant lymphoma. Since the statistics are incomplete, and since ten years is a short time for cancer to develop, it is still too early to tell whether this apparently high rate of cancer is real.

Long-term follow-up of workers exposed to dioxins through industrial accidents is just beginning to yield some evidence. Of 75 workers in a German factory who were exposed to TCDD in 1953 as a result of an explosion in a trichlorophenol reactor, 17 have died, 6 of them of cancer. This is significantly higher than would be expected, particularly as four of the cancers were in the gastrointestinal tract, whereas only one or two would have been expected.

Chapter 11
Reproductive Effects

Halogenated aromatic hydrocarbons are known to interfere with reproduction in a variety of ways. Their effects on hormones may impair the female's ability to conceive and give birth. Some of the chemicals may be mutagenic, causing changes in the genetic material of male or female parents, thus leading to defects in the offspring. They may be teratogenic, crossing the placental barrier in the pregnant female and causing abnormalities in the developing fetus. Many of them are excreted in the mother's milk and thus may constitute an ingredient in the infant's diet.

The ability of halogenated hydrocarbons to interfere with reproductive function is a well-documented problem in many species of animals. Rachel Carson's *Silent Spring* called attention in 1962 to the fact that birds carrying residues of DDT or other pesticides in their bodies lay eggs so fragile that they break before the baby birds can hatch. Although some species have recovered since the banning of DDT in the United States, others, such as bald eagles, ospreys, brown pelicans, and pheasants, may be suffering the same effects because of PCB contamination.

Mammals

Halogenated hydrocarbons also interfere with reproduction in mammals. Fish-eating mammals, which may ingest large

Reproductive Effects

amounts of these chemicals in their diet and concentrate them still further in their own bodies, are especially vulnerable. Dr. Sören Jensen, the Swedish scientist who originally identified PCB residues in wildlife, has made an extensive study of seals in the Baltic Sea, and predicts that they are likely to die out completely because their ability to reproduce has been impaired by chemicals. The Baltic is heavily polluted with PCBs and a variety of pesticides. These chemicals concentrate in the tissues of the herring, the staple of the seal's diet. Dr. Jensen and his co-workers have identified about 130 different chlorinated hydrocarbons in the blubber of Baltic seals, including over 100 PCBs and their metabolites as well as DDT, hexachlorobenzene, dieldrin, and chlordane. The population of grey seals in Swedish territorial waters has declined from over 20,000 in the 1940s to a few thousand today, and the number is still decreasing. The two other species of seals (harbor and ringed) found in the Baltic are also dwindling, with only 200 harbor seals remaining.

Evidence is strong that the decline is due to decreased reproductive ability. Only 27 to 35 percent of mature ringed seal females were found to be pregnant in recent surveys, whereas the rate is 62 to 95 percent in the same species living in less polluted waters. One study found that almost three-quarters of the females that were not pregnant had damaged uteri, making them unlikely to be able to conceive in the future. Nonpregnant seals were also found to have higher levels of DDT and PCBs in their flesh than pregnant seals.

California sea lions have suffered similar population losses, apparently due to a high rate of premature births. High residues of DDT and PCBs in the bodies of the mother sea lions have been blamed for this problem.

A diet of PCB-contaminated fish has been found to interfere with the reproductive ability of mink. The problem was discovered in 1967 when commercial mink ranchers complained of an up to 80 percent mortality rate in the offspring of female mink fed salmon from Lake Michigan. Dr. Jensen experimented with feeding PCBs to pregnant mink at a concentration normally ingested by seals of the Baltic. The average number of mink kits decreased to 2.9 per female from a normal of 5.1. Mink fed higher concentrations did not give birth at all. The fetuses all died and were resorbed by the mother.

Experiments with monkeys

In an attempt to gain evidence more relevant to human beings, scientists at the University of Wisconsin in Madison have studied the effects on monkeys of small amounts of halogenated hydrocarbons in the diet over long periods of time. After consuming 2.5 or 5.0 ppm of PCBs for four months, menstrual cycles of the females showed definite changes, including irregularity, increased duration, and heavier bleeding. After three more months, the monkeys were mated with unexposed males. Of the eight females who had received the heaviest exposure, only one delivered a live infant. Two did not conceive, and five aborted, resorbed the fetus or delivered a stillborn infant. Among the eight monkeys fed 2.5 ppm, five delivered living infants. The six infant monkeys born to the 16 exposed females were all smaller than normal, and they began to show signs of PCB poisoning as they received further exposure through their mothers' milk. These infants developed chloracne, swelling of eyelids, loss of eyelashes, and discoloration of the skin. Three of them died. The remaining three, weaned at four months of age onto a PCB-free diet, were hyperactive and showed deficiencies in learning ability.

Thus, even relatively small amounts of the chemicals severely impaired the ability of the female monkeys to bear live infants and, in addition, seriously damaged the infants, which had been exposed both prenatally and through the mothers' milk. Furthermore, the effect was long-lasting. After weaning, the mothers were placed on a PCB-free diet for a year and then bred again. The new infants, too, showed the effects of the maternal exposure. They were smaller than normal, had some skin pigmentation, and showed deficiencies in the immune system.

The University of Wisconsin researchers also found that dioxin and PBBs had similar effects on monkeys' ability to reproduce, causing failure to conceive, abortion, and stillbirth.

Those infants that were born tended to be smaller than normal. Other researchers have discovered similar problems in dogs exposed to some insecticides, including DDT, aldrin, and dieldrin.

Yusho women

Women and infants exposed to PCBs through the Yusho accident showed many of the same effects as the experimental monkeys. Sixty percent of the women reported abnormalities in their menstrual cycles. Measurements of hormones in their blood and urine found abnormally low levels of estrogens and progesterone. Infants born to mothers who had been exposed to the PCBs during pregnancy were smaller than normal, had a dark brown pigmentation of the entire skin, swelling of the eyelids, abnormal tooth development, and retarded growth. Such infants continued to be born for several years after the mothers' exposure had ceased.

Effects on males

Male reproductive capacity may also be affected by exposure to halogenated hydrocarbons. PCBs and dioxin caused decreases in sperm production in male monkeys. DDT, aldrin, and dieldrin interfered with sexual interest and ability in male dogs. There have been reports of decreased libido among Yusho men, Michigan farmers exposed to PBBs, and farm workers exposed to a mixture of herbicides and pesticides that included dieldrin.

Teratogenicity

Chemicals that cross the placenta from mother to fetus may cause the fetus to develop abnormally. Thalidomide is the most well-known of these so-called teratogens. There is evidence that dioxin may be another. Rats and mice exposed prenatally to TCDD are sometimes born with malformations including kidney abnormalities and cleft palates. There is evidence that the frequency of cleft palate and spina bifida were abnormally high among infants born in Viet Nam during the war years, when herbicide spraying was heaviest. There is also a report of a higher rate of stillbirths at that time in a province that was being heavily sprayed. Concern over the possibility of birth defects led many Italian women exposed to dioxin through the Séveso explosion to seek therapeutic abortions. No increase in birth defects have been reported, however, among the approximately 3000 babies born to women of the Séveso area in the months following the accident.

Hexachlorophene

An alarming report on hexachlorophene as a possible teratogen was presented at the Science Week conference by a Swedish physician, Dr. Hildegard Halling. As chief physician of a hospital for chronic diseases, she had noticed what seemed to be an unusually high rate of birth defects among infants born to her staff. Speculating that this was related to heavy use of hexachlorophene-containing soaps, she examined records of employees of six hospitals and compared the frequency of birth defects in infants born to mothers who used hexachlorophene soap with those who did not. Among the exposed groups, mostly nurses who washed their hands ten to sixty times a day, there were 25 severe malformations in 460 births. No severe defects occurred among 233 births to women who had not used the soaps. Minor abnormalities were also more common in the exposed group compared with the unexposed group.

Dr. Halling's results have been challenged by Swedish medical authorities. Nevertheless, Dr. Carl Keller of the U.S. National Institute of Child Health and Human Development said he believed the evidence was strong enough to recommend that women of child-bearing age not use hexachlorophene. Evidence from animal experiments supports Dr. Halling's conclusions—exposure of female rats to hexachlorophene during pregnancy has resulted in malformed fetuses.

Mutagenicity

Chemicals may also cause birth defects by producing mutations, or changes in the genetic material of the parents before conception occurs. Not all mutations cause obvious defects, but

Reproductive Effects

most are in some way deleterious. They may make us more susceptible to certain diseases, for example. Mutagenic chemicals may thus bring about subtle, long-term damage to the health of the population without being recognized as a hazard. Since many mutagenic chemicals are also carcinogenic, there is ample reason to identify these chemicals and minimize our exposure to them.

Evidence is scarce on the mutagenicity of halogenated hydrocarbons. Most experiments have not found PCBs or PBBs to be mutagenic. Dioxin, however, is suspect. Standard tests for mutagenic activity using bacteria found dioxin to cause mutations in some strains but not in others. In an experiment in which rats were fed dioxin twice weekly, abnormalities in the chromosomes could be seen with a microscope after thirteen weeks. DDT has caused mutations in a variety of test organisms, including rats.

Mothers' milk

Infants of mothers who carry residues of halogenated hydrocarbons in their bodies suffer a triple threat from the chemicals. Beginning with the possibility of inheriting mutated genes, subjected to exposure throughout their development in the womb, they may then ingest additional quantities of chemicals with their mothers' milk. This combination of exposures was fatal to some of the University of Wisconsin monkeys.

Since human milk is high in fat, relatively large amounts of these fat-soluble chemicals dissolve in the milk, much more than in the blood, for example. Consequently, breast-fed infants may receive relatively higher amounts of halogenated hydrocarbons in their diet than any other portion of the population. For example, one study done in 1969 found that samples of human milk from two cities in California contained an average of 60 ppb of PCBs. This meant that breast-fed infants in these cities were getting three times more than the dose considered reasonable by the Federal Drug Administration.

In Japan, one infant who developed symptoms of Yusho disease was exposed solely through its mother's milk. Insecticides, including DDT, dieldrin, and benzene hexachloride, have also been found in human breast milk at concentrations similar to the PCBs, subjecting infants to doses far higher than adults usually receive. The majority of nursing infants in Michigan are being exposed to PBBs. One study found the chemical in the breast milk of 96 percent of women in lower Michigan and 43 percent of those in upper Michigan.

Especial susceptibility of the young

Exposure to even small amounts of these chemicals may be significant when it occurs at such an early age. One study, done at the National Institute of Environmental Health Sciences, found a permanent "feminizing" effect on the livers of male rats whose mothers had been fed small amounts of TCDD and PCBs. When the chemicals were fed to pregnant females, liver enzymes were induced in the newborn offspring. Levels of one enzyme, called UDPGT, are normally higher in adult males than females. But the males exposed to PCBs before birth and through suckling ended up with decreased levels of the enzyme, comparable to adult females.

This finding illustrates a point made by Dr. John A. Moore, also of the National Institute of Environmental Health Sciences, in summing up the Science Week Conference. Exposure to halogenated aromatic hydrocarbons before birth and shortly after, through the mother's milk, can produce irreversible effects that remain long after the chemicals have disappeared from the body. Although most of these chemicals do not produce dramatic malformations, they may cause impairments of function even at low doses. This has been found in enzyme systems that affect metabolism; in behavior, as in the "spinning" mice; and in the immune system, through the infant's vulnerable thymus gland. The first weeks and months of development, when organs are being formed and "learning" to function, is the time when exposure to small amounts of these chemicals is most likely to cause damage.

Part IV
TAKING ACTION

Granted that halogenated aromatic hydrocarbons are harmful to our health, what can be done to minimize our exposure to them? This type of health hazard is unique in that the individual can do so little to protect himself. Usually there is no way even to know whether we are being exposed. For the most part we must rely on the government to regulate by law what exposure will be allowed, to monitor our food, drinking water, and the environment to insure that that exposure is not exceeded, and to oversee the cleanup of contamination where it has occurred.

Chapter 12
Decontamination

The detection of halogenated aromatic hydrocarbons in the environment is a cause for great concern. Since these chemicals are so stable, contamination may persist indefinitely unless something is done about it. Hudson River fish may never be safe to eat in our lifetime if the PCBs are not somehow removed. In Séveso, where the whole area was evacuated because of dioxin contamination, some of the people may never be able to return to their homes. The question of how to remove halogenated hydrocarbons from the environment grows increasingly important as more and more instances of contamination come to light.

Séveso

A variety of ideas were proposed for getting rid of dioxin at Séveso. Attempts were made to infect the soil with a strain of bacteria able to degrade the chemical, but results proved disappointing. The use of solvents to extract dioxin from the soil was considered. A photochemical technique proved promising, in which contaminated surfaces were sprayed with a chemical, ethyl oleate, which enhances the breakdown of dioxin by ultraviolet radiation from the sun.

So far, the major effect has been directed toward removing contaminated soil, vegetation, and animal carcasses and burying them, or storing them in sealed containers until a decision

Decontamination

is reached on disposal. These efforts have been effective enough to allow many of the people to return to their homes. But the most heavily contaminated 200 acres have defied all attempts at cleanup.

The question remains of how to dispose of the contaminated material. Authorities have proposed to incinerate it, but Séveso residents are strongly opposed to the construction of an incinerator in their neighborhood. Most tests indicate that halogenated aromatic hydrocarbons must be heated above 1000°C to be destroyed. This kind of project on such a massive scale has not been tried before, and no one can be certain it will work.

The Hudson River

Scientists debating how to dispose of PCBs from the Hudson River have considered many of the same options. They rejected incineration of contaminated sludge after conducting some pilot tests, which found that burning the material at 1000°C merely drove PCBs out of the residues into the gas stream exiting from the furnace. Treatment in an after burner at 1800°C was necessary to completely destroy the PCBs. The main problem with this method is that it consumes a tremendous amount of fuel—approximately one gallon of oil for every cubic foot of river bottom treated.

PCB-degrading bacteria were also considered for the Hudson River problem. Twenty-two strains of bacteria were found to be able to degrade PCBs. In one test, sludge containing 198 ppm PCBs, injected with a bacterial culture and shaken in a flask for five days, was purified to the degree that no PCBs were detectable. Scientists feel that this method has not yet been sufficiently developed to be practical on a large scale. But it is promising for the future, being environmentally acceptable and potentially inexpensive.

In some waters, authorities believe the best way to deal with PCB contamination is to leave it alone. Since the oils stick to the soil particles and settle to the bottom, it may remain there harmlessly and become naturally encapsulated as uncontaminated soil settles on top of it. This won't work, however, where tides and currents stir up the river bottom, or where dredging is necessary to keep waterways open to shipping, as it is in the Hudson River. For the Hudson, the solution of dredging "hot spots" and burying the sludge in securely encapsulated landfills seems the only practical solution.

The Duwamish and James Rivers

Two other contaminated rivers have served as models for decision-making on the Hudson River problem. In the state of Washington, prompt action after a relatively minor spill of PCBs into the Duwamish River accomplished a 90 percent cleanup. The spill occurred in September 1974, when a transformer being loaded for shipment to Alaska was dropped on the dock and broke open, spilling the transformer fluid into the water. The fluid, 1306 kilograms of PCBs, sank quickly to the bottom.

Cleanup was begun quickly by scuba divers operating hand-held dredges. They removed 418 kilograms of the oil, which was still concentrated in pools on the river bottom. The other two-thirds of the spilled fluid seeped into the sediments and began to move upriver. The Environmental Protection Agency, which was conducting the initial cleanup, recognized that more extensive dredging would be necessary to remove contaminated sediments. By the spring of 1976, the Army Corps of Engineers had been assigned the cleanup task and had devised a procedure for the removal and disposal of the sediments in a nearby storage area. The dredging, which removed more than 730 kilograms of PCBs, was completed in March 1976. Except for the immediate vicinity of the spill site, PCB levels in the river bottom have returned to the previous background values.

In Virginia, pollution of the James River by Kepone is a much bigger problem. The insecticide, manufactured between 1966 and 1975, at the Life Science Plant, a subsidiary of Allied Chemical, in Hopewell, Virginia had been discharged into the atmosphere, into the sewage disposal system, and dumped into clandestine disposal areas. Today, the whole region, including large portions of the James River, is saturated with Kepone. The EPA recently completed a study of this problem and con-

Decontamination

cluded it is probably even more difficult than the Hudson, since the Kepone is not, like the PCBs of the Hudson, concentrated into "hot spots" that could be dredged to remove the majority of the contamination. Complete cleanup by dredging would cost an estimated one billion dollars, plus another three to six billion for treatment of the dredged material. The EPA has recommended that, for the James River, the best course is to proceed with research to look for less expensive methods of cleanup.

Decontaminating people

Although no practical way has been found to decontaminate the James River, the Kepone disaster stimulated a search for a method to decontaminate people who were exposed to the pesticide. The Life Science Plant had closed in 1975 because so many of its employees were suffering severe symptoms of Kepone poisoning. Scientists at the Medical College of Virginia in Richmond found that they could hasten the excretion of Kepone from the body, with a consequent reduction of symptoms, by feeding the workers cholestyramine, a cholesterol-lowering drug. Cholestyramine is a resinous material capable of binding large quantities of Kepone and preventing the pesticide from being reabsorbed from the intestine. The rate of Kepone excretion was doubled in those treated with the drug, the Virginia scientists found, and levels of the pesticide in blood and fat fell twice as fast as in untreated animals or people. Twenty-two former employees of the Life Science Plant were given cholestyramine. Many of them were suffering from severe neurologic disorders such as visual difficulties, tremors, memory lapses, stuttering, and even hallucinations. All were greatly improved after six months of treatment.

The method is now being tested on other halogenated hydrocarbons, and the Virginia scientists believe it may be effective for some of them, including TCDD. It is uncertain whether such treatment would be desirable or effective for most people, who carry only small residues of chemicals in their bodies. But at least for people exposed occupationally or accidentally to large amounts of halogenated hydrocarbons, the treatment offers great hope.

Chapter 13
Identifying the Problems

Governments have often been criticized for not taking adequate action to protect us against the hazards of chemical pollution. Too often a disaster has had to occur before something was done to prevent a repetition. But even with the firmest determination, regulation of these substances is a tremendous task. There are an estimated 63,000 chemicals in use in the United States, and each year approximately 1000 more are introduced. Congress has passed several laws in recent years, vastly increasing Federal authority to regulate the entry of these chemicals into the environment; but the implementation of the laws depends on many factors, not the least of which is determining which substances are, in fact, hazardous to health.

Testing for cancer-causing activity is especially controversial. The most reliable evidence comes from epidemiology—if a well-defined group of people has received a known exposure to a specific chemical, and then suffers an increased rate of cancer, almost everyone will agree that the chemical is carcinogenic. But very few such instances are known, and obviously they cannot be arranged for experimental purposes.

Next most reliable are experiments on animals. The standard protocol for carcinogenicity testing in animals requires at least 500 animals, 3.5 years, and $250,000 per substance. Often at the end of the experiment the results are still controversial. The high dosage levels required to obtain a measurable re-

Identifying the Problems

sponse in a relatively small number of animals is questioned, as is the relevance to humans of results obtained in animals. Much can go wrong during such a lengthy and cumbersome procedure: mistakes can be made, or animals die of other causes, or pathologists disagree on diagnosis. Nevertheless, these experiments most often agree with epidemiologic evidence when it exists, and are generally regarded as the best we can do. Much research is going on, however, searching for cheaper, faster, and more reliable tests. Methods of testing for mutagenesis, teratogenesis (birth defects), neurotoxicity, and immune suppression are also unsatisfactory and in need of improvement.

How to detect the chemicals in food or water is another problem. Usually the quantities of chemicals to be measured are very small, of the order of "parts per million" (ppm) or even "parts per billion" (ppb). This is usually near the limit of what analytic instruments can detect, but they may still be biologically significant. At this level, many things can go wrong with the tests—other substances present in larger quantities may interfere with the measurement, or extra amounts of the suspect chemical from some other source may contaminate the sample in the laboratory. A different question arises as the technology for testing is continually improved. The more sensitive the machinery, the smaller the amounts it can detect of the suspect chemical. Does one ban a chemical at concentrations of "parts per million" or "parts per billion"? What about "parts per trillion"?

Detection and identification become even more difficult when there is only a suspicion that an outbreak of illness among people or animals has been caused by an unknown chemical. In many such disasters of the past, weeks or months went by before the cause of an epidemic was identified. In such cases, the problem may be made almost impossible because analysts have no idea what kind of chemical to look for. In the Michigan disaster, although farmers suspected something was wrong with the cattle feed, test after test found nothing unusual. Only when one chemist accidentally left his mass spectrometer running while he went to lunch, did he find an unfamiliar signal which turned out to be the "fingerprint" for PBBs.

Even when a contaminant is identified, it may turn out not to be the significant one. After the Yusho outbreak, several weeks of detective work pinpointed the PCB-contaminated rice oil as the cause of the trouble. But later studies indicated that the worst symptoms were probably caused not by the PCBs themselves, but by dibenzofurans that were contaminating the PCBs.

Chapter 14
Avoiding Future Contamination

The history of dealing with toxic chemicals has too often been a history of crisis and response. In the 1960s and 1970s, however, increasing public concern led to an increased role of the Federal Government in attempting to prevent disasters before they occur. Since 1970, Congress has passed a series of laws aimed at preventing the entry of harmful substances into the environment: the Clean Air Act, the Safe Drinking Water Act, the Federal Water Pollution Control Act, the Federal Insecticide, Fungicide and Rodenticide Act, the Hazardous Materials Transportation Act, the Occupational Safety and Health Act, the Resource Conservation and Recovery Act (which regulates treatment, storage, and disposal of hazardous wastes), and the Toxic Substances Control Act (designed to identify chemical hazards before they cause trouble).

The National Institute of Environmental Health Sciences was established in 1969 to provide a center for research aimed at identifying harmful man-made agents in the environment and understanding their relationship to human diseases.

The Food and Drug Administration, which has often been unprepared to deal with crises caused by contamination of foods by industrial chemicals, set up in 1971 a Chemical Contaminants Program to prevent or minimize such emergencies in the future. The program is designed to predict and search for chemicals likely to find their way into food, evaluate the hazards, and take corrective action early.

The governmental agency most concerned with protecting the American public against exposure to harmful chemicals is the Environmental Protection Agency. The EPA is responsible for enforcing most of the laws that deal with clean air and water; it is charged with the regulation of pesticides; and it has been given the difficult job of administering the complicated and controversial Toxic Substances Control Act of 1976, familiarly known as TOSCA.

TOSCA

This act, which took six years of negotiation before Congress agreed on an acceptable version, is the first legislation designed to regulate hazardous substances before they reach the environment. It requires industry to notify the EPA in advance of plans to manufacture a new chemical or to put an existing chemical to a new use. The EPA administrator then has broad authority to prohibit or limit manufacture or use of the chemical, or to require special testing. The law also provides for an inventory of existing chemicals and for the testing of those most likely to pose a risk to health or environment. As of the end of 1978, an interagency committee had drawn up a list of 21 suspect chemicals, including five classes of halogenated aromatic hydrocarbons, to be tested for, among other effects, carcinogenicity, mutagenicity, teratogenicity, and environmental consequences.

TOSCA is new and complex, and little has been done so far to implement it except hire staff and begin formulating policies and procedures. Whether this law and the other environmental health legislation will effectively protect our health and environment against halogenated aromatic hydrocarbons and other toxic chemicals depends on the will of the government and of the people. Chemicals are basic to our way of life. Regulating them is controversial and expensive. Economic, political, and social factors must be weighed against considerations of health and safety when decisions are made. These dilemmas were discussed at The New York Academy of Sciences Confer-

Avoiding Future Contamination 67

ence, Public Control of Environmental Health Hazards,* which followed the one on Health Effects of Halogenated Aromatic Hydrocarbons.

But the long-term consequences of our decisions must be considered carefully. The effects of halogenated aromatic hydrocarbons are insidious because the chemicals are so long-lived. A little exposure here, a little pollution there may not hurt. But little by little it adds up. By the time we notice it hurting, it is too late. The person who has been carrying residues of halogenated hydrocarbons in his body for thirty years may develop cancer, or his memory and coordination may begin to fade. The child, exposed in his mother's womb and through his mother's milk, may have learning difficulties because of subtle brain damage, or he may often be sick because of an impaired immune system. But neither will know what caused their problems because they are ordinary problems, and very common.

Like the individual, the whole earth could get sick imperceptibly, until it too is past curing. Cleaning up polluted rivers may cost millions of dollars. Cleaning up the ocean would be impossible. Whole species of life are dying out, some because chemical pollution interferes with their ability to survive or reproduce. Man's own survival depends on the health of the earth. It is in our own best interest not to let it be poisoned.

---

* For the interested reader the subject is considered in detail in a nontechnical volume based on the material presented at this conference which is entitled *Hazards to Your Health: The Problem of Environmental Disease*, by James Gorman, also published by The New York Academy of Sciences.